AGRICULTURAL EDUCATION,

AN INDIVIDUAL, STATE AND NATIONAL NECESSITY;

WITH

Suggestions for the Establishment and Endowment of an Agricultural College in Michigan:

AN ADDRESS

BEFORE THE

Agricultural Society of Calhoun County,

BY

ROBERT F. JOHNSTONE,

Editor of the Michigan Farmer,

DELIVERED AT

MARSHALL, OCTOBER 12, 1854.

"How can he get wisdom that holdeth the plough; that glorieth in the goad; that driveth oxen, and is occupied in their labors, whose talk is of bullocks.—*Eccl* xxxviii: 25."

Published by Order of Calhoun County Agricultural Society.

DETROIT:

PRINTED AT THE DAILY DEMOCRAT ESTABLISHMENT,

Corner of Jefferson Avenue and Shelby Street.

1854.

ADDRESS.

Mr. PRESIDENT, and Members of the Agricultural Society of Calhoun County:

Almost thirty years ago, on the occasion of laying the foundation stone of that mighty column which commemorates one of the most wonderful struggles recorded in history as incident to the birth of nations, the most distinguished statesman of the age, and certainly one of the most illustrious citizens to whom our country has given birth, after declaring that we could not all expect to stand by the side of Solon or Alfred, or compare with Washington and Franklin, and other great men who were their companions in laying the foundations of our government, made these remarks: "There is opened to us a noble pursuit to which the spirit of the times strongly invites us. Our proper business is improvement. In a day of peace let us advance the arts of peace, and the works of peace. Let us develop the resources of our land; call forth its powers, build up its institutions, promote all its great interests, and see whether we, also, in our day and generation may not perform something worthy to be remembered."

Such, nearly thirty years ago, were the words of the great New England statesman.

When they were spoken there were in our State but feeble symptoms of that rapid and vigorous growth which has since characterized western progress and improvement.

The great experiment of canal navigation was untried; and the question as to whether communication by that means could be carried on with profit between distant portions of the country, was yet undecided. The Erie Canal, that artery along which pulsates the life-blood of states and communities, and especially of our own, was not completed. No great line of railroad had been laid down in either the Eastern or Western Hemispheres. The scream of the locomotive had not been heard in the old world or the new. The crossing of the ocean by steam was looked upon as an impossibility; and no electric telegraph had stretched its skeleton wires from one end of the continent to another, transmitting with lightning speed the mystic hieroglpyhs of thought.

Thirty years ago the whole population of the State of Michigan was not equal to that contained in the twenty towns which are comprised in the present county of Calhoun. Thirty years ago the boundaries of Michigan were not established by law, and white settlers in the interior were almost unknown. The beautiful oak openings and fertile prairies that lie around us, now thickly dotted with farms, had not tempted the emigrant so far from the lakes which connected him with the friends he had left behind; and no villages with their churches, their institutions of learning, their busy humming mills, and broad populous streets occupied the places where Marshall, Albion and Battle Creek now stand. No plow had turned a sod on the prairie; no axe had leveled the sturdy growth of your timbered lands. Not even a road had traversed your county; and the venturesome explorer who came out to gratify his curiosity by a view of the interior, trusted himself solely to his compass and the Indian trails, then the only pathways through the wilderness. Such was the condition of this portion of the territory of Michigan, when, before the assembled multitudes on Bunker's Hill, Mr. Webster declared that "Our proper business is improvement."

Let us now, by a few inquiries and investigations, endeavor to ascertain to what extent in "our day and generation" this "business" has been prosecuted; and also trace to their true cause the rapid advances that are being made in the condition of the country and its inhabitants.

Behold, where then was an uncultivated wilderness, where no sound was heard but the cry of the beast of the forest, the song of the wild bird, or the whoop of the Indian hunter, we see now a gathering of intelligent farmers, mechanics, and artists, who have brought hither the productions of their industry and skill, and who seek with friendly emulation to obtain the rewards that shall be adjudged to those who are foremost in the ranks of improvement. Looking upon these evidences of progress, may we not say of those who conquered the wilderness and forced it to yield its fruits to their peaceful arts, that they, in their day, have done "something worthy to be remembered?" and does not such a scene as this bear ample testimony to the foresight, the industry and zeal with which the duties imposed upon them by the "spirit of the times" have been fulfilled?

Within these few years past the face of the country has undergone an almost entire change. Is not this in a great measure to be attributed to the revolutions that have taken place in agriculture as an art? That this is the "noble pursuit" to which, more than to any other, "the spirit of the times" is directing the attention of this generation, all will concede; and a glance at the different principles which have governed the pioneer and his successor will at once exhibit the true secret of our present prosperity and rapid advancement, as contrasted with the comparative snail-like pace of our predecessors.

The great object with the pioneers, was to economise labor and dispense with capital. The principle by which they were guided merely taught them to cut down the forest and till the land in the most primitive manner. Their successors, on the contrary, by the application of capital in accordance with the discoveries of experience reduced to science, are enabled to give increased fertility to lands already subdued, but the natural productiveness of which has been either impaired or exhausted.

The practice of the first is comprised totally in local, traditionary experiences handed down from father to son. It can be reduced to no standard; it is governed by no rules; improvement in it is only the result of necessities which force those who follow it to depart from their usual routine to overcome some new and unexpected difficulty. This was the method of the pioneers. They knew they had nothing to do but to make use of the treasures which lay at the surface. Like the first explorers of gulches and canons of the auriferous Pacific state, they had only to turn up the soil anywhere, and in the most unskilled manner, to obtain the reward of their labor. Their achitecture was as rude as their husbandry. The art of building the log house, the log barn, and brush fence, was not to be learned from works of science. No treatise on architecture ever taught how log was piled upon log, nor did any agricultural economist ever explain the art of making brush fences. The implements they were obliged to use, and the stock they raised were all in keeping with their times and their necessities. Their hands were their capital; their expenditure the sweat of the brow. Their income was not reckoned by dollars and cents, but by the increase of cleared acres, and by the additions made from year to year to the stock which grew up around them. The prices in the markets gave them no concern, for they were too far off to be affected by any fluctuation that might occur. They had enough for their own consumption, and sometimes a surplus to barter away with any new settler who might chance to come among them, for some article which, to the new comer, was a superfluity—to the pioneer a luxury. When a piece of land was worn out, the idea of enriching it from the barn yard never swept across their minds except as a far off vision of something that might be practised where land was dear, and help might be had for the asking. Cattle were seldom yarded, or kept where their manure could be made available even if wanted; and if, by chance, a horse made his nightly lodging in the log stable, the mound that grew up beneath the square hole that served for a window in the rear of his domicil, was undisturbed year after year, except by the weekly addition to its size, and finally descended to the next purchaser of the premises, a grass-grown moument to the unskilled system of the pioneer.

You have all seen these mysterious mounds, they are rich in hidden treasures; some of you perhaps, taught by the cunning arts of chemistry, have forced them to yield back those treasures to the land they had helped to impoverish. But the pioneer had not learned that art. It was easier and cheaper for him to clear off another portion of the forest, or turn over a new sod on the prairie, than to attempt to renovate an old and worn out field.

This is the method of culture still pursued on new lands at a distance from market, where the soil is luxuriantly fertile; producing large crops with but little labor; and such was the condition of farmers here before the introduction of railroads. Corn grew rank but was not worth transportation to market. Wheat was grown, but not in quantities to more than satisfy the demand of the population of the small towns and villages sparsely scattered over the state. Oats were raised, but only to feed the horses that were raised on the farm; and cattle, sheep and wool were cared for only to supply home necessities.

The advent of the locomotive was a signal for, or rather the begining of a revolution for which the pioneers were not prepared. They were shrewd enough to see that they were out of place in the midst of improvements which were being made around them by the introduction and application of capital to a business for success in which they had been taught to rely only on their strong arms and unbending constitutions. They did not choose to become students, or to unlearn what they had been taught; or rather, they had become too old to learn aught that was new in an occupation to which they had been trained from earliest youth. They did not like innovations; they could not endure to see their toilsome, hard-fisted efforts set at naught by the introduction of senseless gearing and machinery; new implements were not to be tolerated in place of the old and tried. They laughed at the idea of Merino sheep being better than the coarse wooled breed they raised, and which were as choice as any at the East when they left that somewhat indefinite region. The hardy, half-wild cattle which came without cost, and required little care in keeping, they considered far more profitable than the high-bred, high-fed animals whose ponderous frames they imagined would consume more food during the winter than their carcases would be worth in the spring.

All were not so; but the class we speak of made up their minds that they could do better by accepting what they considered a fair compensation for their improvements, and moving to some other locality, where they could again commence, and carry on that system of pioneering which required no capital but their sturdy frames, and no learning but that transmitted to them by their fathers, or which they had picked up while conquering some of the difficulties incident to the life of the backwoods pioneer.

No disparagement to this class, as individuals, is meant by alluding to them in this manner. They have had their counterparts in all ages and amongst all nations; but it is a settled fact that wherever their mode of culture has prevailed to such an extent as to become a national characteristic, the result has been the same—where no advances have been made in agriculture, no innovations upon old systems ventured upon by agriculturists, there has been a corresponding national stand-still, or, more properly speaking, a national retrogade movement. For an example or illustration of this you need not look beyond the past unwritten history of your own county.

Think you, that if instead of the tomahawk, the bow and the scalping knife, the dusky aborigines that you have driven from these forests, had possessed the axe, the plow, and the reaper, they would so soon and so easily have yielded to your advances? No, had these broad, fair fields been tilled by them; had their roof-trees been as deeply rooted in the soil as yours are now, they would have bid defiance to your puny attempts at supplanting them, as you have since to their threatened return to their ancient hunting grounds. Their tomahawks and their arrows passed through the air, and left no trace of where they had been; their foot-paths through the unbroken forest, narrow, devious and darkened as their own minds, have disappeared under your furrows, and their very existence is fast becoming a tradition of the past.

It is true that their mode of life was a grade or two below that of the pioneer, but each in its degree produced the same results. They did not "advance the arts of peace," and theirs was not "an age of improvement." They fled at the sound of the pioneer's axe, as that very class of men to which allusion has been made, have voluntarily exiled themselves to still farther western wilds before the scientific innovations which result from the advancement of the arts of peace. Yet the mission of these men was in its degree as noble as yours, as faithfully performed, and as necessary to your comfort and success, as is the proper application of the sciences and systems you have introduced, to the perfect developement of the resources of your land.

All honor to the pioneers of the west! the breaking-up team of the age! still pushing sturdily on in advance of the lighter implements with which science comes to cultivate their rudely broken furrows, to scatter the seed for you with unconscious hands, to gather up the bountiful harvest in her arms of wood and steel, and lay it submissively at your feet! Some who were pioneers are yet amongst you, to witness, to share in, and be benefited by these triumphs of art; but they are, generally speaking, men of the age, men of improvement,—those who are willing not only to advance with the world, but to do all in their power to aid in its advancement.

This, friends of agriculture, is a slight sketch of that system, if system it may be called, which was practiced by the early settlers, which still prevails in many of the partially settled counties, and which will continue to prevail to a greater or less degree, so long as land is cheap, labor scarce, wood an incumbrance, and markets are at a distance.

Now let me give you a few comparative statistics by way of contrast, and to show the changes that have been made in the value of your agricultural products since the construction of railroads. The two great staples of your state are wool and wheat. By examining the census tables of 1840 and 1850, we find that the increase of the number of sheep in the State during those ten years, was 650,000; and the increase of the number of pounds of wool in the same time was 1,789,908. Of wheat in the same period the increase was nearly 3,000,000 bushels.

Look now at the difference made in the value of a single bushel of wheat by the two different modes of transportation. On the ordinary highways of the country, it was seldom that more than fifty bushels of grain were brought to market at a single load; and whatever the distance from market might be, the value of that load was decreased in proportion to the cost of its transportation. Thus, when wheat was worth $1.50 per bushel in Detroit, a load of thirty-three bushels, or one ton, if already in the market, would have brought $49.50, while in the granary of the farmer, thirty miles distant, it would only have been worth the same money minus the expense of marketing, which is estimated by some at the rate of fifteen cents per mile, making the cost of transporting one ton thirty miles, $4.50. This would not be far out of the way, taking into account the wear and tear of the team, harness, wagon and man; especially when it is remembered that as there were then no commissioned agents stationed along the road to receive the grain, and forward it to the great marketing emporiums, the farmer was obliged to take it there himself, and thus risked the temptation to appropriate a part of the money for liquid returns which frequently floated off a goodly portion of the proceeds of the harvest, as, possibly, some of you have had occasion to know!

At the distance Marshall is from Detroit, with which market it is most naturally connected by geographical position, a bushel of wheat was thus reduced in value from $1,50 to ninety-nine cents, and were it worth only seventy cents in Detroit, the farmer of Calhoun county would expect to realize but nineteen cents per bushel for his labor and expense of raising it. So with corn; when it was worth fifty cents per bushel in Detroit, it could not from this distance be marketed there and leave a profit, even though the raising of it cost nothing. But the locomotive, and the iron track, stretching from point to point, across valleys, through hills, and over prairies, has changed all this by bringing the markets of the world to your doo rs. The bushel of wheat which realized to the producer but nineteen cents, became worth the whole seventy, minus only the railroad freightage, which at present is but about eleven cents per bushel, leaving in the hands of the cultivator a fair return for his investments in land and labor.

In regard to live stock the change was still more to the advantage of the farmer. Cattle, which previous to the introduction of railroads would not pay their own transportation, suddenly rose in value. But the enhancement of the value of native stock was not altogether satisfactory. It was soon found that a steer worth 60 or 70 dollars at the Bull's Head in New York or Brighton, cost no more for carriage from the oak openings of Michigan, or the prairies of Indiana or Illinois, than one worth but 35 or 40, and shrewd dealers discovered that they could afford to pay a better price for good animals of improved breeds than they could for the raw-boned, thick-hided, half wild brutes which grew up without care and with but little expense to the owner. The means of rapid communication now afforded, brought western farmers into closer competition with their more advanced brethren at the east, and rendered the introduction of improved breeds of cattle of the utmost importance.

Eastern agriculturists had already taken several decisive steps in the way of progress. Canals and other public works, by increasing the value of their lands, had also added to their taxes; and this, acting as a spur to their ingenuity, led them to seek out a different and more profitable system of culture than that which thelr fathers had pursued.

They had encouraged the invention and use of new implements; they had tried the horse rake and horse hoe, and pronounced them indispensable. New cultivators, scarifiers, improved plows, and various other implements had come into use, and been adopted as necessary to aid in the labors of the farm. They had also learned that much was to be gained by the introduction of breeding cattle superior in all the points which make such animals valuable to the agriculturist; such as early maturity, capacity to put on flesh and fat, hardiness of constitution, and increased size and weight.

All this the agriculturists at the east had learned and commenced to practice upon, and profit by, before the farmers of Michigan had fairly got out of the woods, or opened their eyes to the necessity of any such improvements. And when the snort of the locomotive announced that it had brought in its train, not only the cars laden with new comers seeking western homes, but also the markets of Boston and New York, and Liverpool, and London, close to the barn yards of Michigan farmers, it took some time for the tillers of the soil to realize the new duties it imposed upon them, the new relations it gave them, the new advantages it created for them, and the vast field of enterprise it opened before them for their benefit and enjoyment.

Many, as we have said, could not submit to these changes, and they moved away farther towards the setting sun; but others courageously looked the innovation in the face, and acknowledged the truth of the statement made by the New England sage, that this was a "day of progress, and an age of improvement."

While the emulation created by a nearer proximity to market, as well as to their more advantageously situated neighbors in the eastern States, thus affected the outward condition of the western farmer, it made necessary also a corresponding change in his mental cultivation.

It became evident that an acquaintance with an entire new range of subjects was indispensable. To be a successful agriculturist one must learn things that were never taught or even dreamed of by the plain, straightforward, strong-handed farmers who had hewed their way to a home, axe in hand, and who were still contented wlth the work of the old-fashioned wooden plow, and the equally antique, triangular harrow.

To acquire that knowledge, both mental and physical, which these new relations, new duties and new privileges have made necessary, to give it a proper, practical application to the business of every day life, to make use of it in developing the resources of our land, calling forth its powers, building up its institutions, promoting all its great interests, and making our day and generation one worthy to be remembered, this is what is meant by Agricultural Education.

Agricultural Education! How long is it since such a term came into use?

Let us for a moment take a backward glance at the state of agriculture amongst the earlier nations of the earth, and, observing its effect upon the human race for successive ages, trace the dawning idea of its fuller developtment by education, up to our own times.

We have proof that the ancients had acquired a degree of skill in cultivation, far back in the times when Medea and Bactria, and the plains of

Assyria and Babylonia were clothed with the luxuriant crops which the teeming soil and warm climates of those lands awarded to the husbandman. But the ruins of their mechanical ingenuity, the works constructed to convey the waters of the Tigris and Euphrates into the distant furrows of fields that are now a desolation and a wilderness, are all that remain. Layard speaking of these in the narrative of his travels over that country, says, "the embankments of innumerable canals, long since deserted by their waters, crossed our path, bearing witness to the skill and industry which once made these barren plains one vast garden." Their implements of husbandry were rude, their method of tillage not progressive, and the result is what the traveler sees, an impoverished soil, and a degraded people amidst the ruins of past greatness.

The same, in effect, might be said of that fertile valley through which flows the mysterious and marvelous river, on whose banks stand the pyramids. The Egyptian husbandman knows nothing beyond the mere mechanical performance of those duties which experience taught his forefathers were necessary to obtain the simplest results; and though that land of corn from the fertile Delta to the cataracts of Nubia, has been known for fifty centuries to produce crop after crop, yielding the most ample returns, there has been no improvement in the implements used, the mode of working, whether in sowing, irrigating, harvesting, winnowing or storing the grain its exhaustless soil produces. The same crooked stick is used to scratch the ground now by the fellah, which was used when the foundations of the pyramids and of the sphinx were laid. The same rude *shadouf* and *sakia* which were then employed to irrigate fields on which no drop of rain ever falls, are still in use to raise the water from the Nile to the fields along its banks. In vain has the Assyrian, Persian, Greek, Roman, Saracen and Turk, in turn swept over the land; no progress has been apparent in its agriculture, and national degeneracy has been the inevitable consequence. While art and architectural science stand immortalized by the everlasting pyramids, the wondrous cities and magnificent temples built in honor of the gods, and which, century after century, whether in the full splendor of their pristine beauty, or in a state of grand and solemn ruin, have excited the curiosity and challenged the admiration of the world, we have no evidence that the idea of agricultural education for the purpose of improvement ever entered into the minds of those who, without the labors of the husbandman, could have raised no conquering army, won no victory, celebrated no triumph; and whose career would not have been worth recording, either on the rolls of papyrus which the antiquarian rescues from the catacombs, or with their mysterious hieroglyphs on their more lasting monuments of stone.

Both Jewish and Egyptian Legislators appear to have been governed by the sentiment expressed by the apochryphal writer, when he exclaimed: "How can he get wisdom that holdeth the plow—that glorieth in the goad; that driveth oxen, and is occupied in their labors; whose talk is of bullocks ?"

In that land whose bright galaxy of warriors, statesmen, historians, philosophers, poets, painters and sculptors has for ages out-shone all others, we find nothing on record concerning the condition of agriculture, except the honors which were paid to the divinities who presided over the growth of the field, the orchard and the vineyard. What we learn of the festivals given in honor of the corn goddess Demeter and the wine god Dionysus, is about all that is left us of the manners, practices and customs of the agricultural population of ancient Hellas. Their great festivals, such as the Olympic, Isthmian and Pythian games, at which the whole male population of the several States were often present, made no such display as we here behold. The only distinction which agriculture enjoyed, was the presence of the priestess of Ceres and her attendant virgins. We read of the swift-footed horses of Alcibiades and Hiero, but have no record that the art of breeding those wonderful coursers that won for the victor in the chariot races, the cherished crown which the proudest king deemed it an honor to bear off, was thought worthy of a single inquiry. While the swiftest runner was considered worthy to have his name recorded in letters of gold upon pillars of the marble of Pentelicus, the tiller of the soil was deemed unworthy of public notice by the great mass of the fickle Greeks. Historians, poets, and artists received honors due to their genius, but we read of no exhibition of the fruits of the earth to encourage improvements in agriculture. While the choicest oxen, the finest bulls, and the fattest rams, blazed upon the altars in honor of the gods, no word was ever said as to what reward was due to those whose patient and plodding care had reared and fattened the animals for sacrifice. To those who raised the fleeces and figs of Attica, the wheat of Marathon, the honeyed wealth of Hymettus, no praise was given. The gods fashioned from wood and stone—those forms of classic beauty, shaped by the genius ot a Phidias or a Praxiteles—received divine honors, while the tillers of the soil, the men whose industry furnished the priests with their most valuable offerings, were esteemed but as a portion of the clods which composed the soil, and capable of no improvement. As a nation, the Greeks neglected the true ground-work of permanent national greatness, and the consequence was inevitable. They exist now amongst the ruins of a glorious, but too short-lived past. It was but too common with the ancients to consider the cultivation of the soil an unworthy occupation for those who aspired to defend their country, or to make war upon neighboring nations, and suited only to the capacities of an enslaved and conquered race. As a striking and ready example, we need only cite you to the condition and employments of the Helots amidst their Dorian conquerors.

What we find upon the pages of such writers as Cicero, Cato, Pliny, Varro, Columella and Virgil, relative to agriculture, gives us no reason to suppose that the Romans deemed it an art susceptible of improvement by the education of the agriculturist. The invention of new implements, the adoption of easier methods of cultivation, the better instruction of those who had the charge of landed estates, or the art of improving the breeds of domestic animals seem never to have occupied their thoughts. We know that much of the patriotic virtue for which the earlier Romans were distinguished, is ascribed to the fact that they belonged to a class who tilled their own possessions with almost penurious care; and the names of some of their most illustrious families, as the Fabii, the Lentuli, the Pisones and Cicerones were significant of their origin from amongst the farmers in the neighborhood of the Eternal City. But whatever tendency there might have been towards agricultural improvement, either in the citizens or their government, it was checked in the outset by the heavy drafts made on the population for their wars of conquest, by which freemen were taken from their homes, and in return the victorious legions supplied their places with enslaved prisoners from foreign lands: and from the introduction of that system we may date the commencement of Roman degeneracy.

We have no record of any nation ever having improved its agriculture, or perpetuated its national greatness, while the cultivators of the soil were in a state of ignorance or thraldom. In confirmation of this, I might, if necessary, still farther refer you to the condition of Europe during the dark and middle ages, when the relation of the tiller of the soil to the estate on which he labored, and to its owner, was aptly expressed in the terms by which he was recognised in law; *vassal, villein, thrall, serf, bondman.* During the prevalence of the feudal system, the land-holders by their tyrannical and incessant demands and drafts upon the agricultural population. entirely precluded the possibility of improvement in that class, and agriculture was in its most degraded state. From this condition it did not revive till the necessities of commerce in England gave it an impetus in the reins of Henry VIII., and of Elizabeth, after the civil wars had ceased, and men began to feel that there was some security against the power and grasping avarice of the rapacious nobles, who had heretofore reaped what the husbandman had sown. But it was not till after the middle of the last century, when some progress had been made in the two sciences of Chemistry and Geology that, as an art, agriculture began to show unmistakable signs of that vitality which now pervades it, and which, gaining as it does each day, additional energy and power, promises soon to place it where it belongs—first on the list of sciences, itself the life and mainspring of all others.

Chemistry with her wonderful powers of analysis and comparison, and with such priests at her shrine as Scheely, Priestly, Lavoisier, Galvani, Volta, Berselius, Sir Humphrey Davy, and a host of gifted men of a more recent date, to explain her mysterious processes, has done much, and will still do much more for agriculture. Geology too, with an equally illustrious array of talent to aid in its development has toiled among rocks of error, surmounted Alps of stubborn difficulties, and steadily progressed through strata after strata of ignorance and superstition, till it not only stands of itself a science with its own symbols, and its own language, but it has also lent its aid in laying broad, and deep, and permanent, the foundations of a thorough agricultural practice. While these sciences were being moulded into tangible forms from the chaotic masses of facts which had been discovered and brought into contact from time to time by the intelligent minds of the age, it was hardly possible that agriculture so intimately connected with both should lie dormant. Indeed it was the necessity of infusing new life into the latter that aroused the spirit of inquiry and research which so far perfected the former. The lands worked were yearly becoming less productive, and the prospects were that the crops would soon be insufficient to sustain the increasing population. A resort to science and the application of principles there unfolded, led to experiments which were successful in restoring fertility to the worn out soil. These successes awakened a new interest in the cause of agriculture. It became a study, a fruitful field of new and exciting experiments, which, soon attracted the attention of men of genius, and enlisted their active services in its behalf. As the benefits arising from the adoption of popular improvements became more apparent in the increased fertility of the soil and the abundant harvests, the necessity of making corresponding improvements in the various animals reared for sale or use, began likewise to force itself upon the minds of agriculturists. The art of breeding became a part of their study. Some turned their attention almost wholly to that subject, and became successful breeders of those superior varieties which are now not only spreading over all England, but the United States.

No one can deny that it is to Great Britain we must look for examples of the most successful application of that system of agriculture which is taking advantage of scientific discoveries and mechanical inventions to make the land repay by its increased productiveness, the immense amount of capital expended upon it. The drill system, alternate husbandry or rotation of crops, drainage, the application of special manures, and the use of the sub-soil plow have each and all been carried into practice there on the most extensive scale. Geology, chemistry, zoology, botany, entomology, comparative anatomy, animal and vegetable physiology, hydraulics, mechanics, meteorology, and book-keeping are there each and all called upon to aid in rendering the education of the farmer perfect. They are made to assist him in developing soils and bringing them to their utmost fertility, and also insuring the highest profit, by surveying, draining, manuring, cultivating and securing the crops, with the least possible waste of land, time, labor and capital. They aid him too in breeding, rearing, feeding, managing and marketing stock, as well as in rendering the produce of the dairy of the best quality and the utmost profit.

Novel experiments in various branches of husbandry have recently given a new and enthusiastic impulse to the cause of agriculture in Great Britain. But as yet, as far as the eastern continent is concerned, this enthusiasm is, with few exceptions, limited to that island, and the prosecution of those extraordinary experiments confined to a class who have all the advantage of wealth and leisure. In speaking of these advancements, therefore, it must be remembered how small a part of the world England is, and how many centuries elapsed after her existence as a nation, before even she began to understand where her true strength lay, and to direct her attention to that noble pursuit upon whose proper development is based the strength, and intelligence, and independeuce of any people. Look over the monarchies and despotisms of entire Europe, and you look in vain for anything like general education amongst the tillers of the soil. The very class from which those despotisms draw their life, and their power to oppress, have not sufficient intelligence to maintain their independence should it be granted to them. *Would* this be so, *could* it be so, if they were not degraded almost to a level with the brutes that share their labors? Their manner of cultivation and their implements are sadly at variance with the pretensions to civilization made by the governing class. An American resident in Paris, the very centre of European civilization, writing from that place during the past summer, says; "I sometimes wonder that anything grows in France, the tools used in gardening and in agriculture are so uncouth and unhandy." And another traveler, an accomplished writer, and an experienced and accurate observer, who within a few years past has visited nearly all the countries of Europe for the purpose of noting the improvements in agriculture, lately] declared in an address before the Agricltural Society of New York, that while in one or two localities where schools had been established, much good had been effected, yet, with few exceptions, in spite of an imposing array of institutions and professorships, there was very little improvement in the general agriculture of the country. And this is the testimony of all travelers.

In our own country, we find that political and social institutions make a wide difference in the condition of its farmers compared with those of Europe. The very fact of their self-dependence gives them a shrewdness of discernment and an aptness of application for which the European tenant, who is in reality the practical agriculturist of that country, can never know the necessity. This fact will also contribute greatly to the rapid diffusion of scientific agricultural education, when it is once known that it can be obtained, and to what extent it can be applied with advantage to the business of every day life.

There is a spirit of emulation and inquiry abroad in the land; and this spirit is daily becoming more prevalent and inquisitive, and less easily satisfied with ordinary operations and ordinary results. There are secrets in nature, influencing the success or failure of the farmer which only science can discover; and it has been the study of some of the best minds of the age to investigate the laws by which she governs her most successsul operations; to pry into the secrets of her laboratory, to watch, to experiment, till by patient toil and research they have in a degree become masters of the arts by which she excels. Through the instrumentality of the Press, the fruits of their labors, like the good seed of the thrifty and generous handed husbandman, have been sown broadcast over the land.

See what has been done for agriculture by these means within the last twenty years. In 1835 there was but one periodical devoted solely to its interests in the United States; now they are almost as numerous as the States themselves. One of the first efforts made to arouse the minds of farmers of this coantry to their own interests was that of the few, but influential and persevering men who organized the New York State Agricultural Society in 1838. Those men had observed the good effects of the Royal Agricultural Society of England, and resolved to awaken in their own State and country a spirit of inquiry similar to that which had been aroused by their English prototype. The result of that effort is well known. The first step was taken in the grand march of improvement; State after State came into the ranks; communities and counties marshalled their companies of independent yeomanry and wheeled into line; now, all are actively engaged in battling the common enemy, the allied legions of ignorance and error. At their annual gatherings on public parade grounds like this, they exhibit the trophies of their victories, compare the spoils they have won and incite each other to bolder exploits and nobler efforts in the future. The farmers of twenty years ago enjoyed none of the advantages these social gatherings afford. They had no opportunity of learning what you may learn each year, from the published reports of such societies, in relation to the improvements being made in the methods of tilling the soil; or of examining new implements, or comparing the merits of the various breeds of foreign and domestic animals. They leveled the forest, broke up the sod, and prepared the way for you.

Compare such a scene as you here behold, with those witnessed by the pioneer of twenty years ago, and ask yourselves if, as a county, your progress is in proportion to your increased advantages. Truly, with regard to soil and situation, "the lines have fallen to you in pleasant places; you have a goodly heritage." Nature has done her part, and he who imagines he could improve on her handiwork must think himself a being of superior order. Mechanical art and inventive genius have done their share by bringing the markets of the world to your doors on the iron track, and by giving you tidings from distant lands by the spirit-like rappings of the electric telegraph; implements have been furnished for your hands and thoughts for your brains; have you made them subservient to your interest—which is the interest of agriculture—which is the interest of your country? You have been progressing it is true; you could not well do otherwise with these improved means of communication open to you; but are you prepared to take advantage of all the facilities within your reach, to increase your own knowledge by the addition of valuable and successful experiments, that you may transmit it, thus enriched, to your children?

In the simple art of growing grass, is there one here who feels satisfied that he has arrived at perfection? You think you have done well when you have made your fields average three tons to the acre, but can any of you say there is any reason why, by the proper application of skill and capital, you might not as well average five tons as three? If by some improved system of cultivation, you might increase the products of your grass lands sufficiently to enable you to send from each acre, two head of improved cattle annually to the New York market, instead of the one or half or three quarters of one you now do, would not that, besides being something to your own private advantage, be worth telling of, even in presence of such an assembly as this? I believe that it will be done in a few years; and that such is the spirit of emulation and progress amongst the farmers of Calhoun, it will be done by some of them. You may deem this visionary, but it is no more so than what has already been accomplished by you might have seemed fifteen or twenty years ago. Let your memories run back to that period; think what the ground you stand upon looked like then; and think if, looking into the future then, you could have realized the possibility, in so short a time, of an exhibition like that we have here witnessed to-day! Could you have believed that the wilderness would have been made to yield such superior productions in such abundance, or that you could have made such a goodly show of blooded cattle, horses, sheep and swine; or that it would ever have entered into the minds of men to contrive such implements as you here behold, for the use of agriculturists who had taken up their abode, as it were on the very outskirts of civilization? Where then were the reaping and mowing machines, the cultivators, the plows, the harrows, the threshing machines and horse powers, the construction of which adds so much to the industrial wealth of the country, aside from the advantage their possession affords you by increasing your power of tillage. The idea that such things might be in your day, would have seemed visionary to you then; but believe me, you have not seen the end yet.

Let us now take a brief view of the present condition of agriculture in Calhoun county; indeed, I ought not to confine myself to this county alone, for the farmers in all the southern and central portions of the State are in nearly the same relative position,—all more or less flourishing as they have been able to profit by the advantages which the general progress of the country has placed within their reach. I am tempted here, by way of introduction, to repeat a remark recently made by a friend of mine who is here with you to-day. In speaking of your prosperity now, he exultingly declared that the time had gone by, when to meet the expense of reaping his crops, it frequently happened that the farmer's only resource was a chattel mortgage on his standing grain; and he rejoiced at the idea that such a legal instrument would soon become so obsolete that even the older limbs of the law would scarcely be able to recall the phraseology necessary to make one out!

The times that made such an expedient a necessity to the farmer are indeed past. And now, in passing along your smooth country roads, bounded by neat, well-built fences, or by long lines of walls composed of stones gathered from the adjoining fields, we see those fields almost free from the remains of the original forest, with scarcely a stump to disfigure their green expanse, clothed with waving crops, or luxuriant with rich pasturage, or the cultivated grasses; and when in those pastures we behold flocks whose progenitors came from the royal stocks of Rambouillet or the Escurial, or from the still more distant and finer wooled Saxon races; and when we discover the grazing herds to be of the improved breeds which it has been the care of the wealthy landholders of Great Britain to originate and rear; when orchards of choice fruits greet our eyes, and well arranged yards where cribs, granaries, stables, sheds and barns bear ample testimony to the thrift and economy of the proprietor; and lastly, when we see the residences themselves, in all the elegance of their fair proportions, adorned by the skill and taste of the architect, elaborate with ornament without, and fitted up with the comforts and luxuries of life within—we may well judge that agriculture has outlived the doubtful period of its infancy and is ready to enter upon a new phase of its existence.

This is a fair picture of your prosperity here; and a type of the condition of agriculture generally throughout those portions of the State which have been longest subjected to cultivation. Muscular strength, and what knowledge you possess, have wrought out their good effects; but does not experience teach you that still to advance, more capital must be applied; and that to obtain profitable returns for increased expenditures, more mental energy must be thrown into the occupation of the farmer than has ever yet been devoted to it? That mental energy must be nourished and strengthened by giving to it the knowledge for which it hungers—the learning necessary to its existence; and in order that this may be done, is it not evident to you that we must summon to our aid a thorough practical system of agricultural education? Let that summons be issued by you; see you to it, that it is respected and obeyed; then will the present farming population of Michigan not only prove themselve worthy successors of their honored pioneers, but they will be able also to perform something worthy of being remembered.

We often hear of instances where the best lands of our own State, as well as of others, have deteriorated; where the great staple crops do not grow as luxuriantly as formerly, or average as large a yield of sound grain to the acre. Now, the reflective farmer knows very well the limit to which his own knowledge and experience will lead him in attempts to remedy such an evil, but is not sure how far beyond that he may venture with safety. Progressive in his nature, he regrets his want of that knowledge which might serve as a guide beyond the circle to which he has all his life been confined. It is true he may feel that even if the means to gain it were now at hand, it would be too late for him to profit by them; but he has sons and daughters who will soon occupy his place, and he knows that to be successful in the future, they must be wiser than he has been. Besides these, there are *other* instances where further light and knowledge are required. In close proximity to some of our most thriving towns, and lying around, and among the improved lands, occupying often most valuable locations, are large tracts yet untouched by the hand of labor. Untouched, not because they are valueless; the farmer knows there is wealth under the bogs and standing pools of those wild morasses, but how shall he set about to make it available? He too lacks the knowledge which his brother on the exhausted field is wanting. But where shall they go to obtain that knowledge? You all know it is not to be had in the State; in fact it is not to be had to its full measure in the country. To improve these swamps properly and profitably will require on the part of the agriculturist an acquaintance with practical engineering, and a knowledge of hydraulics, and of surveying applied to the art of draining. He must understand too, the nature of the vegetation which now occupies those lands as well as of that he would cultivate on them, which he cannot do without a thorough knowledge of botany, vegetable physiology, and the principles and application of agricultural chemistry. By these means alone can be made available the fertilizing properties which lie like outspread and untold treasures of ungathered wealth all around us, in every hollow and basin, where they have been accumulating for ages. We would stock the rich fields into which those swamps should be converted, with flocks and herds whose welfare demands that we should be prepared to guard them against disease, and provide for all their wants; to do this, is it not necessary that we should also have the opportunity to learn something of animal physiology and veterinary science and practice? All this knowledge, or the means to obtain it, should be at our command. Those who preceded us did not need it; we have not yet felt the want of it; but does not a wise foresight, a prudent sagacity tell us that the time is approaching—nay, is now at hand—when it will be found indispensable to us, as well as to those who are destined to take our places? And can we be satisfied that we have accomplished all that our position at the present day requires of us, if, *with the means in our hands,* we permit to be neglected the great first duty of furnishing to the rising generation an adequate and fitting opportunity to obtain scientific and practical instruction in all that relates to the theory and practice of agriculture? I say, *with the means in our hands;* for there never was a State more richly endowed with the means of giving to her sons and daughters such instruction, than is, at the present time, the State of Michigan.

Our State needs an Agricultural College. That college should be founded on a basis separate from, and independent of, all others. Such an institution cannot be established without an ample endowment; and what more ample endowment could be desired than the gift which the United States Government has already made to Michigan of the swamp lands within her borders? This grant is made with the simple understanding that it is to enable the State to construct the levees and drains necessary to reclaim those lands and render them susceptible of cultivation. Five million two hundred and seventy-six thousand acres have already been confirmed to the State; and more are still to be set off to her. Of this immense territory, as yet, but one hundred and forty-four thousand eight hundred and fifteen acres have been sold, amounting to the sum of $114,607—$53,597.26 of which has been paid into the Treasury, leaving, outstanding, but drawing interest, some $61,000. For this money the State has no immediate pressing use. Should the whole of the lands be sold at the same rates as those already disposed of, there would be a fund of over four millions of dollars at the disposal of the State. Surely nothing could be more reasonable than that a portion of this immense fund should be set apart for the purposes of education, when that education is designed to be of such a nature as would tend to increase the value of those very lands which furnish the endowment of the institution where it is obtained. Would it not be better, by some organization on the part of the farmers of the State, thus to devote a part of the proceeds of these land sales to the increase and diffusion of knowledge adapted to their own necessities, than to have it lie in the State Treasury, a bait and a temptation to greedy and grasping political adventurers? And are the farmers of the State not entitled to at least four or five per cent. out of this swamp land fund, for the purposes of agricultural education, when the very investment of this per centage will, in a few brief years, double the principal from which it may be taken?

The establishment of an institution that would supply our present wants, ought not to cost over $50,000, and need not, if managed with that economy which a prudent individual of good judgment would deem necessary in his own business. Give it a farm of a thousand acres, give it all the necessary buildings, and the professors competent to teach all that is required, and is it too much to say, that after the first initiatory years, it would annually send out at least two hundred graduates, well versed and skilled in all the arts that would enable them sucessfully to cultivate whatever lands might come into their own possession; capable men, also, into whose hands might be entrusted the drainage of those very swamps which the State, by its acceptance of the gift from the General Government, has engaged to improve.

I believe the whole annual cost of an institution like this, for salaries, professors, and apparatus, after it reached its highest possible developement, would not be over twelve or fifteen thousand dollars. Should the

State refuse to grant a sufficiently large appropriation, the balance might easily be made up by stock subscriptions from the friends of the pupils and of the institution. In any event the necessity for such a college is beyond all question,—I say let the experiment be tried. To make it answer the end in view, it is of vital importance that the school system and discipline should be thoroughly practical. The students ought to be trained in a course of actual practice, in the management of tools, implements, machines, animals and land, in the tillage of fields, the application of manures, the sowing of seed, the raising of crops, getting them ready for market, and their sale; and also in the raising, fattening and marketing of stock. No hired help should be employed to do any part of even the most toilsome and most disagreeable labors; but the several classes should feel a pride in endeavoring to excel each other in the management of the sections committed to their charge.

In the improvement of land by draining, the students should not only be taught the art of surveying, levelling and laying out the drains, but with pick and spade in hand, let them open the trenches, lay the tiles, and perfect the work. If horse power is used, let them learn the art of managing the machines. They have a right to this knowledge, and should not submit to be defrauded of it, as too many have been.

I do not believe in that species of agricultural education which is taught only by lectures, or by study in the closet. What can the in-door student know of the qualities of soils, the lay of the land, the nature of animals, or of the various expedients to which the fluctuations of the weather may drive the quick-witted and active out-of-door farmer? or last, and not least, what would such a student know of the market value of his crops, or of the state of the markets themselves? It is as preposterous to say that a farmer can be made without the practice of his business, as that a divine can be made without the study of the Scriptures. The military establishment at West Point is a standing comment on the folly of the course that has heretofore been pursued in our academies and institutions of learning in regard to agricultural education. There, no cadet is considered worthy of promotion to the most trivial command in the regular army, till he has been well drilled in the performances of all the duties of the common soldier. Does not common sense teach us, that at least equal care and pains should be taken with the education of those who form a much more numerous and necessary class of community, and upon whose success so much of the happiness of life depends? Hitherto what has been taught of agriculture, has generally been made to come in after the regular course of study, as though it were only of third or fourth rate importance; indeed, as it was only in theory at last, it seemed to make very little difference in the end whether it was studied at all or not.

Agriculture requires, and deserves, and calls for an institution *of its own.* Let it have one! It is in your power, farmers of Michigan, to establish one which will be an honor to yourselves and your State. Your educational funds have already been too long, and too exclusively devoted to fitting out students for the learned professions, theology, medicine, and jurisprudence. It is time there should be a change. I would cast no reflections on the learning, talent and skill engaged in these professions; nor do I ask you to pull them down. That is not necessary. They are institutions built up by time and popular opinion, and are subject to them; and whilst they deserve to be sustained, they will be, and with no grudging hand. But it is for you to rear another, not to say more honorable, but of as great utility, and more peculiarly your own. The State has neglected your interests in this respect; no care has been taken to aid in your advancement; but now the means are in your power; use them in such a manner as to elevate your calling in your own eyes, and the world will soon learn to do it honor. These hints are not thrown out as though such an establishment as I speak of, were some visionary project of an imaginative brain—some far-off, impracticable theory for you to speculate upon; I hope so see it built in reality, and these suggestions are intended for practical use. They are prompted by the necessities of the times. Farmers of America must keep pace with other branches of industry and art, if they would not have their nation add another to the list of those who have flourished for a season on the fat of the land, run a career of transitory splendor, and having robbed the earth of its power to sustain them, withered away like a tree without root, leaving only the stark and blighted skeleton of what they might have been, a lasting memorial of their own folly. And thinking-men are waking up to these necessities: from every part of the Union they are calling for light and knowledge on the subject of agriculture. The idea that farmers must be *educated for their occupation* is everywhere rapidly gaining ground; and in uttering these sentiments, and urging this matter upon your immediate consideration, I am but giving expression here to what is becoming the popular opinion of the times.

Members of the Society, and farmers of Calhoun! Your position is an enviable one! Situated near the centre of a State which is rapidly developing its wealth and its capacities; with a climate sufficiently severe in the extremes of its summer suns and wintry frosts to call forth your energies in counteracting its effects; connected, directly, at all seasons with the east, and its emporiums of export and import; with a great iron thoroughfare in your midst, along which pours, in a ceaseless tide, the emigration and commerce of the growing west, and which the future promises will, at no distant day, be the means of connecting you with the regions that border on that mighty ocean whose limitless expanse is but just beginning to be whitened with the sails of commercial enterprise; with States and communities advancing, and great national interests arising on all sides; you cannot stand still even if you would. Bear in mind then your obligations to those who have gone before you, the duties you owe to yourselves and your successors. The experience of all ages tells you that where the mental cultivation of the tiller of the soil has been neglected—where labor has not been honored—there have the people and the land together sunk in the scale of civilization and importance. Wherever men have held in contempt that great law which the Omnipotent gave to the first born of the earth when he declared: "In the sweat of thy face shalt thou eat bread;" there have the nations fallen; and no wealth or splendor has sufficed to save from desolation and ruin, the land where the obedient subjects of that law have been despised and dishonored. While the earlier nations respected that divine command, they increased in strength and might and majesty, till their glory became the pride of mankind! Their rulers and their men of power neglected to honor it; and the very grass refuses to grow on the mouldering ruins, the deserted plains and barren deserts, which now mark where they once flourished in all their pomp and magnificence. The plains of Shinar—the great valley of the Nile—the hills and vales of classic Greece—the slopes of Sicily, once the granary of the mistress of the world—the pontine marshes and maremna of Italia, all speak to us from the darkened past of the consequences of contempt for that eternal principle by which our tenure of the soil—fertile as when God gave it to us—is continued from age to age; and Europe as she is stands a living and present witness that there the mandate has not been, and is not yet respected! Why else do her oppressed and starving crowds rise up and seek our shores? and from what other cause springs that constant fear of dissolution which makes her banded despotisms tremble? Upon the progress and prosperity of agriculture depends the permanence of all other progress. Arts, sciences and commerce may all reach a certain point, may all live and thrive for a time on the stimulus accorded them by despotic power—that ability to trample labor under foot—to repress the longing and earnest desire for knowledge in those who live by it, by condemning them to hopeless and helpless ignorance; but the reaction comes—the "baseless fabric" of their unsubstantial greatness falls—corruption and desolation are their end. But with an agricultural population whose "business is improvement," no tyranny can ever thrive; liberal institutions must prevail, and freedom of speech and action be the inheritance of the tillers of the soil.

An old Greek myth represents that from a serpent's teeth, sown by the Theban lawgiver on the soil he had selected for his home, sprang the men who built and guarded his city ; so from the teeth you shall wrench from the dragon of ignorance will spring the men who will guard the institutions you bequeath to them.

Men of Michigan ! to you is committed the sacred trust of educating your youth to fit them to perpetuate unimpaired that prosperity of which your fathers and yourselves have laid the foundation. The General Government has placed in your hands the means of establishing a noble institution, which, if wisely conducted, may be the pioneer and prototype of many others which shall spring up in different and distant parts of this wide spread Union. You have it in your power to elevate that noble pursuit in which you are engaged ; to extend its usefulness ; to add new treasures to the general wealth; to give salubrity to localities now deserted, and to increase the productiveness of the land you have chosen for your homes. From your midst go forth the young men whose enterprise and talents and genius build up our cities, and add to the wealth and renown of our State in mercantile and professional life ; let them go with minds imbued with respect for your occupation, and a love for the rural scenes amidst which they were nurtured. To you again will return the physician from his office, the divine from his desk, the judge from his bench, the merchant from his counting-room, bearing the wealth which has repaid years of arduous toil, the reward of industry, to expend in the quiet of country life. Make your homes and neighborhood attractive to men of mind and capital, and the professions themselves, by their wealth and influence and intelligence, shall aid in advancing your interests.

Farmers, the destiny of the State rests with you! Place Michigan where she belongs, in front rank with the most advanced of her sister States in that noble pursuit to which the spirit of the times so strongly invites; then shall our freedom as a people be perpetuated, the integrity of our institutions be sustained, all our great interests promoted; and we, too, shall leave behind us a monument more lasting than one of brass or stone—a name and memory to be honored to the latest generation.

5

www.ingramcontent.com/pod-product-compliance
Lightning Source LLC
LaVergne TN
LVHW011143110826
845150LV00008B/2480
* 9 7 8 1 4 1 8 1 9 2 9 2 1 *